MOTORCYCLES
A GUIDE TO THE WORLD'S BEST BIKES™

HARLEY-DAVIDSON
AN ALL-AMERICAN LEGEND

rosen publishing's
rosen central®

NEW YORK

For James and Cindy

Published in 2014 by The Rosen Publishing Group, Inc.
29 East 21st Street, New York, NY 10010

Library of Congress Cataloging-in-Publication Data

Roza, Greg, author.
Harley-Davidson: an all-American legend/Greg Roza. — First edition.
    pages cm. — (Motorcycles: a guide to the world's best bikes)
Audience: 5-8.
Includes bibliographical references and index.
ISBN 978-1-4777-1855-1 (library binding) — ISBN 978-1-4777-1872-8 (pbk.) —
ISBN 978-1-4777-1873-5 (6-pack)
1. Harley-Davidson motorcycle—Juvenile literature. 2. Motorcycles—Juvenile literature. I. Title.
TL448.H3R69 2014
629.227'5—dc23
                                                                                      2013013198

Manufactured in the United States of America

CPSIA Compliance Information: Batch #W14YA: For further information, contact Rosen Publishing, New York, New York, at 1-800-237-9932.

CONTENTS

INTRODUCTION

The year 1903 was a notable one for emerging transportation technology in the United States. The Wright brothers tested the first controllable, powered aircraft; the Ford Motor Company released the first Model A; and the first Harley-Davidson motorcycle was produced. The fledgling motorcycle company—first located in a shed in Milwaukee, Wisconsin—was started by brothers William and Arthur Davidson. They were soon joined by brother Walter Davidson and friend William Harley.

The first bike the Harley-Davidson Motor Company made in 1903 was not a particularly great success, but neither was it a failure. The four founding members, who were trained engineers and bicycle makers, learned a lot from the experiment.

In 1905, the first official Harley-Davidson motorcycle hit the streets with a bigger motor and a sturdier frame. By 1906, production rose from eight motorcycles to fifty. William Harley and the Davidson brothers implemented other changes to increase interest in their product, including a new fork, improved suspension, and, of course, more power. That year, the company also hired workers, provided specialized training, and started setting up a network of national dealerships. These early models were called singles, but they earned the nickname the "Silent Gray Fellows."

In 1907, Walter Davidson entered a two-day, 400-mile (644-kilometer) endurance race from Chicago, Illinois, to Kokomo, Indiana, and back. Walter, and his 1907 Silent Gray Fellow, was one of three competitors to finish the race. It was a pivotal moment in the young company's history. In the next few decades, hard work and endurance became key concepts in the Harley-Davidson factory, which had far outgrown the original shed. An American icon had emerged.

By 1920, the Harley-Davidson company had the largest motorcycle factory in the world, and it produced twenty-eight thousand bikes that year. By 1921, Harley-Davidson was selling bikes in sixty-seven countries through two thousand dealers. The company continues to evolve, and its motorcycles still wow fans and break records. In 2004, William Harley, Arthur Davidson, Walter Davidson, and William Davidson were inducted into the U.S. Department of Labor's Hall of Honor. This is a showcase for Americans and American companies who have made significant contributions to the field of labor and improved overall life in the country.

The Harley-Davidson Motor Company has become a cultural icon in the United States and around the world. The motorcycles in this resource are just a few of the many great machines that have come out of Milwaukee. However, these "hogs," as the motorcycles are sometimes called, truly help demonstrate why Harley-Davidson has remained in the spotlight for over 110 years.

Each year, Harley fans from all over Europe travel to Spain to enjoy Barcelona Harley Days with other Harley fans.

CHAPTER ONE

1911 MODEL 7D

After its initial notoriety, Harley Davidson strove to improve its motorcycle designs. One crucial development was the more powerful V-twin engine. A V-twin has two cylinders at an angle to each other, forming a "V" shape. Two cylinders means more cylinder space, which means more power. Although Harley-Davidson wasn't the first motorcycle company to experiment with V-twins, it has been creating superior V-twins throughout its history. The V-twin's low, throaty rumble has become a Harley-Davidson trademark (although early Harleys came with mufflers).

The first V-twin was not a success. In fact, it was a dud! However, it proved to be an important step in the process of mastering the production of high-performance V-twin motorcycles.

Early Trials

In 1908, Walter Davidson tested an early V-twin in a two-day road race from the Catskill Mountains of New York to Long Island. Walter earned a perfect score and beat sixty-one other competitors. Although the company was not officially involved with racing, these testing ground races helped Harley-Davidson fine-tune its products, and they also proved

Serial Number One, the oldest existing Harley, was built in 1903 or 1904. Note that this early model has just one cylinder.

that Harley-Davidson could produce a motorcycle worthy of endurance racing.

In 1909, the Harley-Davidson company released its first commercial V-twin motorcycle as part of its Model 5 series. The Model 5D was more powerful than the single-cylinder engines. The Model 5 was a single-cylinder engine with a displacement of 30 cubic inches (494 cubic centimeters) that produced 4.3 horsepower. The Model 5D had an engine displacement of 54 ci (880 cc) and produced 6 horsepower. It could reach 60 miles per hour (97 km per hour), thanks to the two cylinders aligned in a 45-degree angle.

However, the 1909 V-twin had some major flaws. Like many early motorcycles, the 5D used a leather belt to transfer power from the engine to the rear wheel. The 5D lacked a way to control the tension in the belt, which meant the belt was prone to slipping. The engine intake valves were atmospheric, meaning they relied on suction to open and close. However, this setup did not work well with the V-twin design. The 5D was difficult to start, and it was no faster than the single-cylinder Model 5s.

Despite an early V-twin racing victory, the Model 5D wasn't a big success. In fact, Harley-Davidson records show that the company recalled all twenty-seven Model 5Ds. This demonstrates the company's dedication to perfection, not to mention customer satisfaction.

The Second Try

After taking a few years to improve the design, Harley-Davidson released the Model 7D V-twin. The 7D had a "magneto ignition" starter. That means riders had to use their legs to turn pedals, exactly like those found on a bicycle, to start the engine. Pedaling allowed the driver to accelerate before the engine took over.

The 7D was a 45-degree-angle V-twin with 49.48 cubic inch (810 cc) engine displacement. Although smaller than other V-twin motorcycles of the era, the 7D produced 6.5 horsepower. The single-cylinder motorcycles of the time

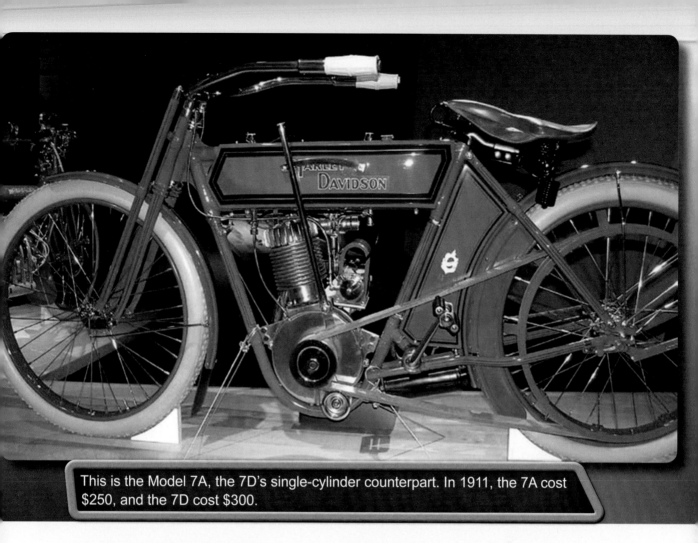

This is the Model 7A, the 7D's single-cylinder counterpart. In 1911, the 7A cost $250, and the 7D cost $300.

traveled about 30 miles per hour (48.2 km/h). The 7D consistently reached 60 miles per hour (96.5 km/h). More fins were added to the engine to help air cool the cylinders and control engine temperature.

The company spared no expense to fix the problems of the earlier model. The 7D came with a drive belt tensioner lever to counteract the 5D's belt-slipping problem. It also featured mechanical valves instead of the automatic valves of the 5D. Mechanical valves are powered by a valve train, which ensures that the valves open and close in a regular pattern,

IMPROVING UPON SUCCESS

In 1912 Harley-Davidson released the Model 8 series to compete in the young sport of motorcycle racing. An updated version of the 7D called Model X8E had a 61 ci (989 cc) engine to create even more power—a total of 8 horsepower. It was capable of reaching 65 miles per hour (104.6 km/hr) and higher. The X8E still had a magneto ignition, but now a chain replaced the belt to deliver power to the rear wheel. Chains eradicated the slipping problem present with leather belts.

The X8E was Harley-Davidson's first motorcycle with a clutch ("X" in the name stood for clutch). A clutch allows the driver to control the transmission of motion from the engine to the wheels. The X8E had a left-side lever that controlled a rear hub clutch. When engaged, it allowed the rear wheel to spin freely without being powered by the engine. This meant the driver didn't have to stop and restart the engine every time he or she came to a stop. It also made starting out much smoother.

The Model 8s were streamlined, sturdier versions of the Model 7s. Sprung suspension was added below the seat to make the ride even smoother. The seat was set lower, which improved stability. Wheel skirts were added to help keep mud off of the engine. The improvements resulted in several notable wins for Harley-Davidson V-twins, including a 1912 San Jose road race where the X8E raced 17 miles (27.4 km) ahead of its rival motorcycles. After 1913, the majority of Harley-Davidson bikes would be V-twins.

rather than relying on suction to open the valves. This allowed the 7D engine to rev higher and create more power.

Harley-Davidson's new V-twin was an F-head engine, sometimes called an IOE (intake over exhaust) engine. In this type of engine, the intake valves are positioned in the cylinder head, above the cylinder itself. The exhaust valves are part of the engine block below the cylinder. Harley-Davidson kept using the F-head engine design with great success until 1929.

Built to Race

The Model 7D had 28-inch (71.1-cm) wheels and a 56.5-inch (143.5-cm) wheelbase. Its gas tank held 2.5 gallons (9.46 liters) of fuel. The 7D's steel frame was sturdier than previous models. It featured sprung forks (tubes with springs inside them) for increased suspension and a lower center of gravity. The belt tensioner allowed the driver to maintain engine power, especially on hill climbs, which aided greatly in endurance challenges. These changes made the 7D even more attractive to racers.

In 1911, there were more than 150 other motorcycle companies in the world. The competition was steep, but the men at Harley-Davidson were up to the challenge. The 7D was a popular bike, and it proved its value on the racetrack and during endurance races. Most 7Ds were driven by private owners. Thanks to their success, Harley-Davidson officially opened a racing department in 1914. Today, the 7D is a rare find. You might find a restored 7D in a private collection, although you can see a Model 5D at the Harley-Davidson Museum in Milwaukee, Wisconsin.

CHAPTER TWO

1936 MODEL EL "KNUCKLEHEAD"

By the late 1920s, the Harley-Davidson factory in Milwaukee was the largest of its kind in the world. It was still run by the original founders, but now their sons helped. The Great Depression made life difficult for many Americans and U.S. businesses. The Harley-Davidson Motor Company weathered the Great Depression by selling generators based on its motorcycle engines. It also reached a wider market by releasing a three-wheeled service motorcycle called the Servi-Car, which stayed in production until 1973.

Even as the country struggled to crawl out of the Great Depression, Harley-Davidson searched for new ways to compete with its rivals. The question, as usual, was: how do we create more power? The company decided to stop manufacturing single-engine machines altogether, and it never looked back. However, it also sought to revolutionize its V-twin designs and create a more powerful, more stylish motorcycle. A major engine redesign led to one of the most iconic Harley-Davidsons of all time.

What a Knucklehead!

Like many American businesses during the 1930s, Harley-Davidson was forced to reduce its staff because of economic

The Knucklehead, with its all-American ruggedness, helped shape the iconic look that we still see in Harley-Davidson models today.

difficulties. However, this didn't stop William Harley and his team of engineers from forging ahead on new designs. They started from scratch to develop a new, groundbreaking product. Between 1931 and 1935, they struggled with a new lubrication system and persistent oil leaks. This issue actually continued to be a problem, but the company would eventually work it out. It was an era of redesign and innovation, and the results were nearly perfect.

The bike of the 1936 E series was the result of many years of hard work. It was often called the 61 E for its 60 ci (988 cc) engine. Other times it was called the 61 OHV, which is short for overhead valve. The EL had higher compression than the E. There was even a sidecar model, the ES. Fans nicknamed it the "Knucklehead." This name comes from the shape of the valve covers, also called rocker boxes. The E series, and especially the EL, would prove to be a highly popular model that set many standards for the company.

OHV V-Twin

The 61 EL was the first Harley to feature an OHV 45-degree V-twin engine design. In the years leading up 1936, most Harley-Davidson motorcycles came with a flathead, or side valve, engine. In this design, the valves are in the engine block rather than the head. The valves are positioned in a chamber next to the cylinder. The engine gets its name from its shape—the head is flat with an opening for the spark plug, which is placed directly over the combustion chamber. The camshaft directly powers the valves, which is one of the noted benefits of the design.

In an OHV engine, the valves are in the cylinder head. The camshaft is in the engine block below the valves, so rockers are used to transfer movement from the camshaft to the valves. The rockers tip back and forth, and they're covered

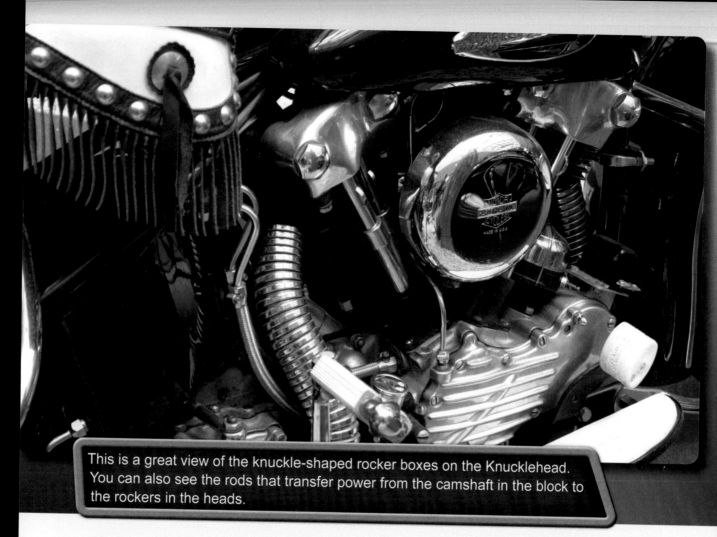

This is a great view of the knuckle-shaped rocker boxes on the Knucklehead. You can also see the rods that transfer power from the camshaft in the block to the rockers in the heads.

by the knuckle-shaped valve covers, or rocker boxes. The 61 EL's engine had a single camshaft, which proved to be more efficient and quieter than previous models.

Big Twin

The 61 EL is considered a big twin, which means it was in the larger class of V-twin for its time. It was twice as powerful as its flathead predecessors. The cylinder heads had a hemi-spherical shape to allow larger, stronger valves and higher compression. The Knucklehead's pistons had a shorter stroke than previous flathead models, which allowed it to rev higher and harder.

The 61 EL big twin could create 40 horsepower at 4,000 rpm. Factory reports claimed a top speed of 100 miles per hour (160 km/h), although some featured a speedometer that went up to a whopping 120 miles per hour (193 km/h). The 61

1941 FL

In 1941, the company released the F and FL versions of the Knucklehead. The FL Knucklehead retained many of the EL's features. In fact, the two motorcycles were practically identical.

As with the E and EL models, the FL version had higher compression than the F version. The FL had a 74 ci (1,208 cc) engine displacement. Other updates were made to help accommodate the larger engine. This included a stronger crankcase, more efficient rocker arms, an improved carburetor, and larger intake ports. The improvements led to 48 horsepower at 5,000 rpm and a top speed of 95 miles per hour (152.8 km/h).

A poor lubrication system plagued the 61 EL, so the FL was given an enclosed, recirculating lubrication system. An impeller (a propeller enclosed within a tube) increased the pressure in the oil system, ensuring that oil reached all parts of the engine evenly. Despite the efforts to perfect the Knucklehead's oil system, leaks would prove to be its greatest flaw. This was an issue addressed by the motorcycle's replacement in 1948—the FL "Panhead."

EL's top speed was generally around 90 miles per hour (144 km/h), which became one of its most attractive features to buyers.

The Knucklehead was Harley-Davidson's first four-speed motorcycle. The transmission and clutch were very dependable. The motorcycle featured drum brakes and a chain drive. The streamlined fenders, large seat, and iconic valve covers made a big impression on buyers. Art-deco styling was popular at the time, and it fit this bike perfectly.

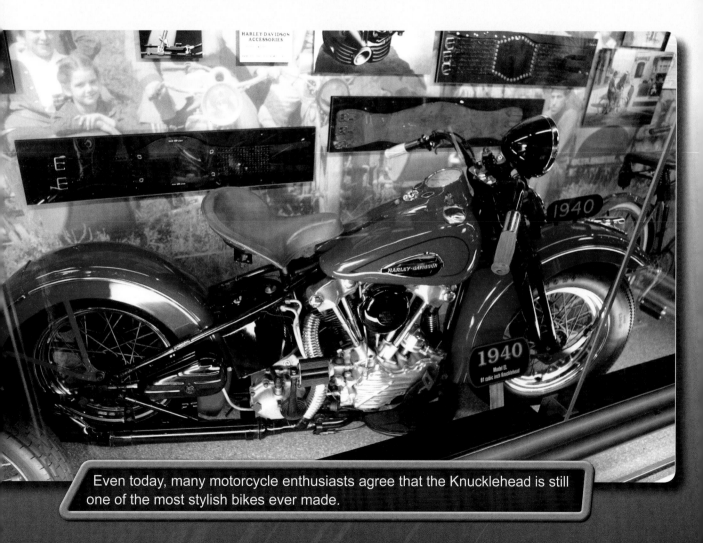

Even today, many motorcycle enthusiasts agree that the Knucklehead is still one of the most stylish bikes ever made.

Something for Everyone

The results of the new 61 ci (989 cc) engine were even better than the company had hoped for. In fact, the E series had something for everyone. Immediately, the Knucklehead caused a stir by breaking two amazing records. Long-distance rider Fred Hamm set a record by driving his Knucklehead 1,825 miles (2,937 km) in twenty-four hours. Rider Joe Petrali set a land speed record by reaching 136.183 miles per hour (219.165 km/h) on a Knucklehead. Knuckleheads were also very good in hill-climbing events. In addition to being a big

Motorcycle racing legend Joe Petrali competes in the 1936 American Motorcycle Association's national hill-climbing event, which he won riding a Harley for the second year in a row.

hit with racers, the 61 EL quickly became a favorite of police forces struggling to keep up with newer, faster automobiles. Many other riders purchased the Knucklehead simply for its good looks.

The 61 EL heralded a turn of fortune for the American manufacturer. In 1937, Harley-Davidson sold over eleven thousand bikes, a total significantly better than previous years. Although sales of the Knucklehead were weak during the United States' involvement in World War II, it proved to be one of the most popular Harley-Davidson models of all time. In 1947, the final year of Knucklehead production, Harley-Davidson sold more than ten thousand EL and FL motorcycles, which amounted to about half of its sales that year. Many fans were sad to see it go, so it's no wonder that collectors still love these motorcycles today.

CHAPTER THREE

1941 WLA

The United States entered World War II in 1941. That year, Harley-Davidson halted production on most domestically sold motorcycles; very few Knuckleheads were produced during the United States' involvement in World War II. Instead, Harley-Davidson focused its efforts on aiding America's military. It did this by producing an iconic military machine: the WLA.

The WLA had its start in 1937 with the WLDR, a civilian racing bike. It was designed to compete with Indian's Sport Scout on racetracks and on the road. The WLDR used a long-standing tradition at Harley-Davidson: the 45-degree flathead V-twin engine. With an engine displacement of 45.54 ci (746.33 cc), it was smaller than the big twin, 61 EL, but it cranked out 27 horsepower. Expert motorcycle tuners modified the WLDR to get better performance from it. Aluminum cylinder heads helped keep engine temperature down. Larger intake ports allowed more air to enter the cylinder and increase horsepower.

In 1941, Harley-Davidson released a factory racing version of the WLDR based on these changes. It did very well against the Indian Scout in competition, just as the company had hoped. It was a durable, dependable machine that could reach 85 miles per hour (137 km/h).

Building an Army Bike

In the years leading up to the United States' involvement in World War II, the government tested motorcycles from several domestic manufacturers in search of the perfect wartime vehicle. It wasn't looking for a combat motorcycle. Rather, it wanted a dependable workhorse for scouting, policing, and delivery duties. The army specifically requested a motorcycle that wouldn't overheat at low speeds and reached no more than 65 miles per hour (105 km/h).

To accommodate the army's requests, Harley-Davidson made several key adjustments to the WLDR. It reduced the

There are many resources available for people interested in restoring an old WLA to its original glory.

displacement of the flathead 45-degree V-twin to 45.12 ci (739.38 cc). This is known as detuning an engine. Instead of tuning the engine for optimal performance, it can be detuned to provide more economical results. Detuning the flathead resulted in lower compression in the cylinder, producing 23.5 horsepower at 4,600 rpm. The WLA came with a plaque warning operators not to exceed 65 miles per hour—which was in fact the motorcycle's top speed. Once the WLA was loaded up with equipment, including a 40-pound (18.1 kg) radio, the machine's top speed dropped to about 50 miles per hour (80 km/h).

The chain-driven WLA was designed to be the perfect military motorcycle. The lower displacement put less strain on the engine, allowing it to conserve fuel and last longer. Designers gave it the Knucklehead's new recirculating oil system and a more dependable oil pump. It had a sturdy three-speed gearbox and clutch. Along with the cylinder heads and larger intake valves, these upgrades allowed the WLA to run consistently at low speeds for long distances.

Harley-Davidson also added several features to make it combat ready. The army requested a bash plate underneath the bike to protect it from rocks and other debris. The WLA also came with a new tubular front fork that was several inches longer to increase ground clearance. The normally loud, rumbling Harley V-twin was muffled by a fishtail exhaust pipe. The space over the rear wheel had a flat surface large enough to hold a bulky military radio. Two large saddlebags flanked the rear wheel, providing plenty of storage. One or

THE 1942 XA

At the start of the war, U.S. army officers convinced Harley-Davidson to make a motorcycle that would perform well in desert climates, such as those in North Africa. They believed the answer was a flat-twin, shaft-driven motorcycle similar to the one used by German forces during the war, which was produced by BMW.

The XA's two cylinders were set 180-degrees apart and stuck out perpendicular to the length of the bike. This design provided far more efficient engine cooling. It also allowed for an enclosed shaft drive. The army hoped it would hold up better than the open chain drive of the WLA in sandy environments. The XA could also be fitted with larger "balloon" tires to generate more traction on uneven terrain and desert sand. The XA was also Harley-Davidson's first foot-shift-controlled transmission.

Even after insisting on the manufacture of the XA, the U.S. military purchased only about 1,100 XAs, and they never saw active service. After the XA experiment, workers at Harley-Davidson toyed with the idea of producing a civilian version of the XA. Despite the many things they learned about the flat-twin engine and shaft drive designs, the company chose to return to more tried-and-true V-twin machines.

more gun scabbards were added to the front, which could also be used to carry a hand tool, such as a shovel.

Uncle Sam Wants Harley-Davidson

In March 1940, the government ordered 745 WLAs. The A stood for "army." Although the company sold very few machines to civilians during the war, the sale of WLAs to militaries around the world gave the company its highest sales since before the Great Depression. Harley-Davidson produced a version called the WLC for the Canadian Army. It also sold WLAs to Great Britain, South Africa, China, and Russia (where a sidecar version of the WLA was popular).

Two other companies produced motorcycles for the government during World War II, but the WLA was produced in far greater numbers. In 1943 alone, Harley-Davidson produced a total of 29,621 motorcycles; 24,717 of them were WLAs, and another 2,647 were WLCs. In total, Harley-Davidson produced 88,000 motorcycles for military use in just five years.

The WLA was a common sight during World War II. It introduced the Harley-Davidson company to many people overseas who had never heard of it. It also became the motorcycle that many GIs learned to ride, and many wanted their own Harley upon returning home. Some people called the WLA the "Liberator." This comes from the presence of WLAs as U.S. forces helped liberate people from the Nazis. In 1943 and 1945, Harley-Davidson was given the U.S. Army-Navy "E" award for excellence in production.

Unlike the WLA, it's harder to find an XA today. This one can be found in the National War and Remembrance Museum in Overloon, Netherlands.

After the War

Harley-Davidson was left with a surplus of WLAs after the war. In 1945, when peace seemed evident, the government cancelled an order for eleven thousand WLAs. However, some estimates say there were also enough spare parts to build another thirty thousand bikes. Many of these were sold to Russia. Many of them found their way onto the streets of America.

With the abundance of surplus bikes and parts, it was possible to buy a "brand new" WLA well into the 1950s, and at a very low price. Ex-GIs and civilians alike enjoyed restoring old WLAs. Keep your eyes open for this U.S military icon the next time you attend a car or motorcycle show. The WLA certainly wasn't as flashy as the Knucklehead, but it has become a beloved symbol of U.S. military excellence, at home and around the world.

CHAPTER FOUR

1972 XR750

Harley-Davidson has had a long, storied racing tradition at many levels, but it's at the 750 cc (45 ci) level where it's truly left its mark. The WLDR engine design of the 1930s lasted into the 1950s, when it was replaced by the KR racer. The KR was a major AMA Grand National Championship contender throughout the 1950s and 1960s. The AMA was the premier motorcycle racing series in the United States, and companies from all over the globe entered racers. Harley-Davidson's KR racers dominated several categories, including dirt track and road racing.

Class C was originally created to feature factory-made, street-legal motorcycles. AMA rules allowed 750 flathead engines, but OHV engines could only have a top displacement of 30 ci (500 cc). This allowed the American-made flatheads to dominate less powerful OHV engines. However, in 1969, the AMA changed the rules to allow any 750 engine to compete in class C. The OHVs from Honda and Triumph quickly proved to be superior to the KR.

Without a backup plan, Harley-Davidson had to adapt quickly. After some scrambling, it produced what many believe is the best bike ever to compete in dirt-track racing.

1970 XR750

In 1970, under pressure to stay relevant in America's foremost motorcycle racing series, Harley-Davidson produced the XR750. However, it was really just a collection of old parts. It featured the KR chassis and an XL Sportster engine. The Sportster was first manufactured in 1957, and it's still popular today. The Sportster's OHV engine made it the first choice when rushing to put together a machine capable of competing with British and Japanese motorcycles.

The Sportster engine was known as the "Ironhead" because it had cast-iron heads and cylinders. It would remain

Around two hundred of the XR750 Ironhead were made, but only about 50 still exist today.

the Sportster's engine until 1987, when it was replaced by the Evolution engine. The Sportster's engine displacement was 53.9 ci (883 cc), which is too high to compete in a 46 ci (750 cc) division. Harley-Davidson destroked the engine. It used a smaller crankshaft, which shortened the distance the pistons traveled in the cylinders. This resulted in a lower displacement and less power. The first XR produced 62 horsepower at 6,200 rpm. Highly praised in the 1960s by fans and experts alike, the Sportster was a fast bike, but it wasn't built for racing. First off, it was too heavy. More problematic, the cast-iron heads overheated frequently. That year proved to be Harley-Davidson's worst season ever, with many of the XRs dying before they hit the finish line. In the off-season, the Harley-Davidson race team worked hard to make the Ironhead engine work, but in the end it gave up and started from scratch.

Do Over

Over the next year, the company developed a completely new OHV engine with new materials, and the XR750 was reborn in 1972. Gone were the cast-iron heads, replaced by lighter parts made of an aluminum alloy. The aluminum heads were bolted straight through the crankcase for added strength. These changes allowed the engine to rev higher for longer periods without overheating. Many people refer to it as the "alloy XR," as opposed to the Ironhead.

Harley-Davidson gave the XR an improved combustion chamber and a one-piece crankshaft. The XR had the first

EVEL KNIEVEL

When some people think of the XR750, they think of American daredevil Evel Knievel. Born Robert Craig Knievel, Knievel was one of the greatest stuntmen ever to live. For a while, his name was synonymous with Harley-Davidson.

The XR750 wasn't Knievel's first motorcycle. Prior to 1971, he had used Hondas, Triumphs, and an Italian motorcycle named American Eagle. Knievel began using the Ironhead XR750 to make long-distance jumps in 1971. On February 28, 1971, he used an XR to jump nineteen cars.

Daredevil Evel Knievel rides a XR750 Ironhead during a stunt show in 1971.

Knievel continued using the XR750 for most of the rest of his career. He used alloy XRs to jump cars, vans, trucks, buses, and even rattlesnakes! In 1977, he planned to use an XR to jump sharks, but he crashed during a practice jump. This would be Evel Knievel's last jump on an XR, and the last jump of his career.

Today, you can view one of Evel's XR750s at the Smithsonian Institution and another at the Harley-Davidson Museum in Milwaukee.

over-square engine. That means the bore—or the distance from one side of a cylinder to the other, also called diameter—is larger than the stroke, or the total distance the piston moves. A shorter piston stroke means less friction and stress on the crankshaft. It also means higher revs. With a 3.125-inch (7.9-cm) bore and a 2.98-inch (7.5-cm) stroke, the XR's displacement was up to 750 cc.

The XR's horsepower is variable, depending on how a race team adjusts or adapts the machine. Initially, the XR produced 80 horsepower at 8,000 rpm. Although the design has changed very little over the years, peak output for the XR is around 95 horsepower at 7,800 rpm, and it can reach approximately 130 miles per hour (209 km/h). The XR often exceeds 100 miles per hour (161 km/h) on the turns alone!

Look the Part

The XR750 has received very few changes over the years, which is part of the attraction for fans and race teams alike.

It had four speeds, a foot pedal to change gears, and a dependable chain drive. For suspension, it had telescopic front forks and rear hydraulic shocks.

Road racing legend Cal Rayborn drives an XR750 to victory during the 1972 Transatlantic Series. This was a rare road race victory for a dirt track motorcycle.

The machine's distinctive look included dual exhaust pipes and twin carburetors. The exhaust pipes came out of the front of the engine, which put the hottest area forward for improved air cooling. The pipes were positioned high and out of the way so riders could lean into the turns. The pipes ended in megaphone mufflers that gave the XR a distinct sound. Two 1.41-inch (36-mm) carburetors with oversized air filters stuck out of the right side of the engine. These carburetors mixed air and gas quickly and increased the amount of fuel that entered the combustion chambers.

The XR was light—as low as 290 pounds (132 kg). A lightweight engine helped make this possible, but many parts were sacrificed to bring the weight down even more. It had a small 2.5-gallon (9.5-l) gas tank and not much of a seat; riders more often used the foot pegs for support and control. Front brakes aren't safe on dirt track racers that need to slide through the turns, so they were left off as well. The rear wheel could be fitted with disc brakes, although even they were optional.

Dirt Track Legend

In early 1972, great American road racer Cal Rayborn stunned the racing world when he took an Ironhead XR to Europe and dominated the Transatlantic Series. Later that year, Rayborn drove the new alloy XR to an AMA Grand National road race victory. Despite early successes in road racing, the XR was destined for greatness in another field.

In dirt track racing, riders steer powerful motorcycles around a 1-mile (1.6-km) or .5-mile (.8-km) oval track covered with loose dirt. Riders speed up through the straightaways, then lean, inner foot on the track, and slide through the turns. Many riders have said the XR is perfectly made for this racing medium.

Not only did the XR750 put Harley-Davidson back on top of the AMA, it made sure the company would stay there for years to come. In more than forty years of AMA dirt track racing, the XR has claimed thirty-two championships. The AMA has even handicapped the XR; under AMA rules, the XR must have 1.2-inch (32-mm) restrictor plates, which reduce the amount of fuel and gas entering the combustion chamber, thus reducing horsepower. This hasn't stopped the XR from coming out on top.

CHAPTER FIVE

2002 VRSCA V-ROD

Simplicity and tradition are two concepts at the heart of the Harley-Davidson company. It has crafted many ground-breaking products, but it has long been known to be conservative when it comes to changes. The 45-degree V-twin engine design has become a trademark of Harley-Davidson. There's no doubt about it: the company has had great success with traditional designs.

Harley-Davidson would turn this notion of tradition on its head with the introduction of the VRSCA V-Rod. In 2002, this unique bike, sporting a high-tech, nontraditional design, stunned the motorcycle world. This quickly turned to amazement as fans and experts alike found out what the new bike could do.

The Harley-Davidson Revolution engine proved to be one of its finest creations, and several models based around it have hit the market since 2002. VRSC stands for V-twin, racing, street, custom. The letter A is used to designate the first of its line.

A Racing Bike at Heart

The VRSCA was inspired by the VR1000 superbike, released in 1994. It was the first motorcycle Harley-Davidson created purely for racing, rather than modifying an existing factory product for that purpose. In superbike racing, two- and four-cylinder bikes race on an asphalt track at speeds of over 200 miles per hour (322 km/h).

The VR1000 was a dual overhead cam (DOHC), 60-degree V-twin. It had four valves per cylinder, electronic fuel injection, and a liquid cooling system. All of these were firsts for Harley-Davidson. The high-tech engine produced 140 horsepower at 10,400 rpm. This power was needed for the speed of super-bike racing.

Harley-Davidson was hoping to create another XR750 to rule the superbike world. In the end, however, the VR1000 never lived up to its hopes. The VR1000 racing program was cut in 2001. But that wasn't the last time Harley-Davidson fans would see the 60-degree V-twin. In fact, the VRSCA V-Rod hit the streets the very next year.

You might be wondering why the VRSCA was such a big deal when the VR1000, with a very similar engine, had debuted in 1994. The VR1000 was a high-power racing bike. Harley-Davidson sold very few of them to the public, and those it did sell cost close to $50,000. So, when the VRSCA V-Rod went on the market at $17,000, many fans jumped at the chance to buy a high-performance motorcycle based on a superbike racer.

On top of this, the VRSCA was vastly different from any commercial bike the company had ever produced. Harley-Davidson, so accustomed to building flatheads and 45-degree V-twins, had come up with something completely new. Jaws

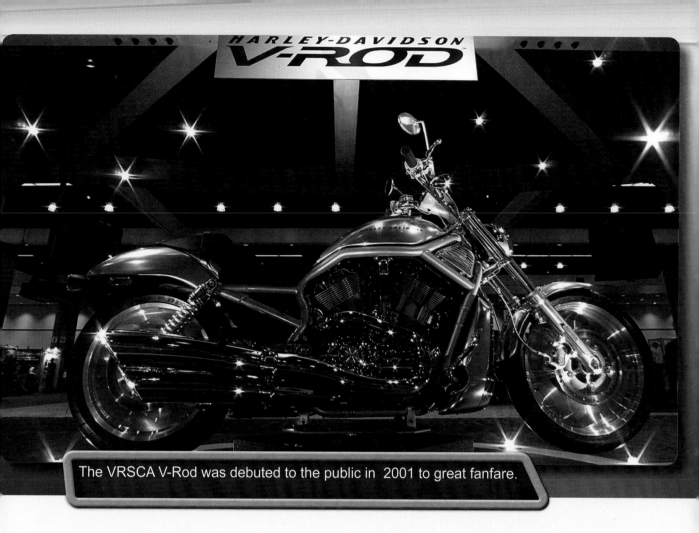

The VRSCA V-Rod was debuted to the public in 2001 to great fanfare.

dropped all over the world when the VRSCA finally rolled off the assembly line.

Motorcycle Revolution

So much about the VRSCA was different from the designs that came before it. The engine alone was a major change of direction. The VRSCA's double overhead cam, 60-degree V-twin engine was appropriately named the Revolution. Compared to the VR1000, the VRSCA "only" produced 115 horsepower at

BACK TO RACING

The VRSCA and the Revolution engine may have been inspired by a racing machine, but it didn't take Harley-Davidson long to take the VRSCA full circle, back to racing. In 2002, Harley-Davidson teamed up with the highly successful Vance and Hines Motorsports team to compete in NHRA (National Hot Rod Association) Pro Stock Motorcycle drag racing. The displacement of the brand-new Revolution engine was increased to 74 ci (1200 cc) to compete with other high-end drag racers, including the four-cylinder Suzukis that had been dominating the event for years. The new drag racing V-rod had a custom frame specifically made for drag racing.

For their first season, the Screamin' Eagle/Vance and Hines team showed moments of greatness, but it proved to be a year devoted to development. The team would do much better in 2003, making a Pro Stock final for the first time since 1980. In 2004, the team won the NHRA Pro Stock Motorcycle Championship. They also set the national Pro Stock Motorcycle elapsed time (ET) record of 7.016 seconds. The team only got better after that! Since 2003, the Screamin' Eagle/Vance and Hines team has won 7 Pro Stock Motorcycle championships. In 2012, the team once again set the ET Pro Stock ET record at 6.728 seconds.

After a stunning 2012 season, during which the Harley-Davidson team won fifteen out of sixteen events, the NHRA decided to change the rules. Four-valve cylinder heads were no longer allowed in Pro Stock drag racing. It was back to the drawing board for Harley-Davidson yet again.

8,250 rpm, but this made it the most powerful street bike the company had ever released.

The VRSCA had an over-square engine, with a 3.94-inch (10-cm) bore and a 2.83-inch (7.2-cm) stroke. This allowed it to rev up to about 9,000 rpms. Engine displacement was 1,130 cc (69 ci). Harley-Davidson gave the VRSCA V-Rod four valves per cylinder. Two intake valves helped move more air and fuel into the combustion chamber. Two exhaust valves helped expel the waste gases more efficiently. To help move all of those valves, the engine required four cams, two for each head, for a total of four cams.

An electric fuel injection system was designed to deliver the perfect amount of air and fuel to the pistons at just the right time. The VRSCA's engine was liquid-cooled, powered by a water pump positioned in the V between the cylinders. Out of all the changes, perhaps this one surprised Harley-Davidson fans the most.

A salesperson poses with a newly sold VR1000 in 1997. This machine was usually more expensive overseas, sometimes more than twice the factory price, due to import fees.

A New Kind of Cruiser

The engine wasn't the only area to receive major changes. The VRSCA was a cruiser, however, and it had some new design features as well. With a 26-inch (66-cm) seat height, the VRSCA was low to the ground, creating a lower center of gravity and greater stability. Harley-Davidson chose 60 degrees

People attending the 2008 Motorrad Messe (Motorcycle Expedition) in Leipzig, Germany, admire the sleek style of a Harley-Davidson Night Rod Special. The Revolution engine in this machine produces 125 horsepower.

for the V-twin because it contributed to the low center of gravity. The 38-degree fork angle was bigger than any other Harley, giving the machine a 67.5-inch (171.5-cm) wheelbase.

The gas tank was moved below the seat for the same reason. The front of the VRSCA looks like a gas tank, but it's a fake! Instead, it hides a downdraft induction system, which helps channel a greater amount of air into the engine. Brushed aluminum styling completed the VRSCA package, resulting in the most radical machine Harley-Davidson had ever created.

The VRSCA was discontinued in 2006, but that doesn't mean you can't buy a V-Rod today. The Revolution engine—now tuned to an engine displacement of 1,200 cc (73.2 ci)—can be found in several newer models, including the Night Rod and the V-Rod Muscle.

SPECIFICATIONS CHART

1911 MODEL 7D	
speed	60 miles per hour (97 km/h)
horsepower	6.5
transmission	direct belt drive (one speed)
fuel capacity	2.5 gallons (9.5 l)

1936 61 EL "KNUCKLEHEAD"	
speed	90 mph (145 km/h)
horsepower	40 hp at 4,800 rpm
transmission	4 speed
fuel capacity	3.75 gallons (14.2 l)

1941 WLA	
speed	65 mph (105 km/h)
horsepower	23.5 hp at 4,600 rpm
transmission	3 speed
fuel capacity	3.375 gallons (12.8 l)

1972 XR750	
speed	130 miles per hour (209 km/h)
horsepower	95 hp at 7,800 rpm
transmission	4 speed
fuel capacity	2.5 gallons (9.5 l)

2002 VRSCA V-ROD	
speed	137 mph (220 km/h)
horsepower	115 hp at 8,250 rpm
transmission	5 speed
fuel capacity	3.7 gallons (14 l)

GLOSSARY

CAMSHAFT A long bar with cams, or cogs, set to open and close the valves as the bar rotates.

CARBURETOR A part that mixes air with fuel mist for an internal combustion machine.

CYLINDER The circular space in which the pistons move, sometimes called the combustion chamber.

ENGINE BLOCK The main part of the engine that houses the cylinders.

ENGINE DISPLACEMENT The volume of an engine's cylinders, which is an indication of power and speed.

EXHAUST The system through which waste gases from a combustion engine are expelled.

HEAD The top, removable part of an engine that covers the cylinders.

HORSEPOWER A unit of power equal to 33,000 foot-pounds per second, or 745.7 watts.

IGNITION The component that ignites the fuel in an engine on a motorcycle or other vehicle.

RPM Revolutions per minute; this refers to the speed at which an engine is operating.

TRANSMISSION A system of gears used to transfer power from the engine to a drive shaft.

VALVE A cylindrical device in an engine that controls the passage of fuel.

FOR MORE INFORMATION

Canadian Motorcycle Drag Racing Association
10757-180 Street
Edmonton, AB T5S 1G6
Canada
(877) 580-9008
Web site: http://www.cmdra.com
This association oversees the motorcycle drag racing circuit in
  Canada.

Harley-Davidson Museum
400 W. Canal Street
Milwaukee, WI 53201
(877) HD-MUSEUM (436-8738)
Web site: http://www.harley-davidson.com/en_US/Content
  /Pages/HD_Museum/museum.html
Explore the history of America's most iconic motorcycle manufacturer and see numerous examples of its finest creations
  at this museum.

Web Sites

Due to the changing nature of Internet links, Rosen Publishing
has developed an online list of Web sites related to the subject
of this book. This site is updated regularly. Please use this link
to access the list:

http://www.rosenlinks.com/MOTO/Harl

FOR FURTHER READING

Barnes, Pete. *Harley and the Davidsons: Motorcycle Legends*. Madison, WI: Wisconsin Historical Society Press, 2007.

Cotter, Tom. *The Harley in the Barn: More Great Tales of Motorcycle Archaeology*. Minneapolis, MN: Motorbooks, 2012.

Cotter, Tom. *The Vincent in the Barn: Great Stories of Motorcycle Archaeology*. Minneapolis, MN: MBI Publishing, 2009.

Gingerelli, Dain, and David Blattel. *Art of the Harley-Davidson Motorcycle*. Minneapolis, MN: Motorbooks, 2011.

Gingerelli, Dain. *Harley-Davidson Museum Masterpieces*. Minneapolis, MN: MBI Publishing, 2010.

Henshaw, Peter, Ian Kerr, and Garry Stuart. *The Encyclopedia of the Harley-Davidson*. Edison, NJ: Chartwell, 2010.

Leffingwell, Randy, and Darwin Holmstrom. *The Harley-Davidson Motor Co.: Archive Collection*. Minneapolis, MN: Motor, 2011.

McDiarmid, Mac. *The Ultimate Harley-Davidson: A Comprehensive Encyclopedia of America's Dream Machine: Developments, Specifications and Design History with 570 Photographs*. Wigston, Leicestershire, England: Lorenz, 2012.

Mitchel, Doug. *Harley-Davidson*. St. Paul, MI: MBI Publishing, 2007.

Newkirk, John J. *The Old Man and the Harley: A Ride Through Our Father's America*. Nashville, TN: Thomas Nelson, 2008.

Smedman, Lisa. *From Boneshakers to Choppers: The Rip-Roaring History of Motorcycles*. Toronto, ON, Canada: Annick Press, 2007.

Tooth, Phillip, and Jean-Pierre Pradères. *The Art of the Racing Motorcycle: 100 Years of Designing for Speed*. New York, NY: Universe, 2011.

BIBLIOGRAPHY

Auto Editors of Consumer Guide. "Classic Motorcycles." HowStuffWorks.com. September 28, 2007. Retrieved March 3, 2013 (http://auto.howstuffworks.com/classic-motorcycles3.htm).

Demortier, Cyril. "2006 Harley-Davidson VRSCA V Rod Review." Topspeed.com. May 21, 2007. Retrieved March 12, 2013 (http://www.topspeed.com/motorcycles/motorcycle-reviews /harley-davidson/2006-harley-davidson-vrsca-v-rod-ar3108. html).

Gingerelli, Dain. *Harley-Davidson Museum Masterpieces*. Minneapolis, MN: MBI Publishing, 2010.

Henshaw, Peter, and Ian Kerr. *The Encyclopedia of the Harley-Davidson*. Edison, NJ: Chartwell Books, 2005.

McDiarmid, Mac. *The Ultimate Harley-Davidson*. London, England: Anness Publishing, 2010.

Mitchel, Doug. *Harley-Davidson*. St. Paul, MI: MBI Publishing, 2007.

Motorcycle Online Staff. "First Ride: 2002 Harley-Davidson VRSCA V-Rod." March 15, 2002. Retrieved March 12, 2013 (http://www.motorcycle.com/manufacturer/harley-davidson /first-ride-2002-harleydavidson-vrsca-vrod-13245.html).

Rafferty, Tod. *The Complete Harley-Davidson*. Osceola, WI: Motorbooks International Publishers, 1997.

U.S. Department of Labor. "Hall of Honor Inductee." Retrieved March 3, 2013 (http://www.dol.gov/oasam/programs/hallof-honor/2004_davidson.htm#.UPSIACfLSak).

Wagner, Herbert. *At the Creation: Myth, Reality, and the Origin of the Harley-Davidson Motorcycle, 1901-1909*. Madison, WI: Wisconsin Historical Society Press, 2003.

INDEX

ABOUT THE AUTHOR

Greg Roza has been writing and editing educational materials for thirteen years. He graduated from the State University of New York at Fredonia with a master's of English in 1997. Roza lives in Hamburg, New York, with his wife and three children.

PHOTO CREDITS

Cover, p. 1 (motorcyclists) Jim Pruitt/Shutterstock.com; cover, p. 1 (sky) photolinc/Shutterstock.com; p. 5 Alejandro Garcia /EPA/Landov; p. 7 Tom Uhlenbrock/MCT/Landov; p. 9 Daniel Hartwig/Wikimedia Commons/File:1911 Harley-Davidson Model 7D (3) - The Art of the Motorcycle - Memphis.jpg/CC BY 2.0; p. 13 Richard Spiegelman; p. 15 Jean-Luc 2005 at de.wikipedia /Wikimedia Commons/File:Harley035.jpg/CC BY-SA 3.0; p. 17 Mark Wilson/Getty Images; p. 18 RacingOne/ISC Archives /Getty Images; p. 21 Joe Songer/AL.COM/Landov; p. 25 AlfvanBeem/Wikimedia Commons/File:G-585 Harley-Davidson XA, 751cc, 23hp.JPG/CC0 1.0; p. 28 Robyn Beck/AFP/Getty Images; p. 30 Ralph Crane/Time & Life Pictures/Getty Images; p. 32 Manchester Daily Express/SSPL/Getty Images; pp. 37, 39, 40 © AP Images; interior pages background elements Dudarev Mikhail/Shutterstock.com, Yuriy_fx/Shutterstock.com; back cover © iStockphoto.com/JordiDelgado.

Designer: Brian Garvey; Editor: Nicholas Croce; Photo Researcher: Amy Feinberg